Is Google's VEO-2 a Game Changer or Just Hype?

Inside the Breakthrough That Could Forever Redefine Text-to-Video Generation

Joe E. Grayson

Copyright © 2024 Joe E. Grayson, All rights reserved.

No part of this publication may be reproduced, distributed, or transmitted in any form or by any means, including photocopying, recording, or other electronic or mechanical methods, without the prior written permission of the publisher, except in the case of brief quotations embodied in critical reviews and certain other noncommercial uses permitted by copyright law.

Table of Contents

Table of Contents

Introduction

Chapter 1: The AI Revolution and the Need for Innovation

Chapter 2: Unveiling VEO-2

Chapter 3: Physics Simulation Excellence

Chapter 4: Creative Capabilities of VEO-2

Chapter 5: Understanding the Underlying Technology

Chapter 6: The Industry Impact of VEO-2

Chapter 7: Challenges and Limitations

Chapter 8: The Future of Text-to-Video Generation

Conclusion

Introduction

Artificial intelligence has become a cornerstone of innovation, driving advancements in industries ranging from healthcare to entertainment. Amid this transformative wave, VEO-2, Google's latest text-to-video generation model, stands out as a groundbreaking achievement. Its release marks a pivotal moment, not just for Google, but for the AI industry as a whole. At a time when competition among tech giants is fiercer than ever, VEO-2 has emerged as a symbol of technological excellence, pushing the boundaries of what AI can achieve. This model doesn't merely aim to improve video generation—it seeks to redefine it entirely, setting new standards for creativity, precision, and realism.

Google's journey in the AI space has been marked by both triumphs and challenges. Once seen as the leader in AI innovation, Google faced stiff competition in recent years as rivals like OpenAI and Meta introduced models that captured global attention. OpenAI's GPT models revolutionized natural language processing, while Meta's generative systems carved out niches in image and video creation. As these companies gained momentum, Google appeared to be lagging, fueling speculation that the tech giant was losing its edge in AI development. However, December brought a dramatic shift in this narrative. With VEO-2, Google not only reclaimed its place at the forefront of AI innovation but also delivered a model that surprised even the most skeptical industry watchers.

This book embarks on an in-depth exploration of VEO-2, examining its revolutionary capabilities and its potential to reshape the AI landscape. From its unparalleled physics simulations to its ability to generate videos with unprecedented realism and coherence, VEO-2 demonstrates why it matters in today's competitive environment. We will unravel the story of how this model surpassed industry benchmarks, outperformed its competitors, and reignited Google's status as a leader in artificial intelligence. Through this analysis, we aim to provide readers with a comprehensive understanding of what makes VEO-2 a game changer and why its impact extends far beyond the technology itself.

Chapter 1: The AI Revolution and the Need for Innovation

The state of artificial intelligence before VEO-2's release was both dynamic and fiercely competitive. AI had evolved into a race for dominance, with companies striving to outpace one another in creating models that pushed the limits of generative capabilities. Text-to-video generation, a field once considered unattainable due to its complexity, became a battleground for innovation. Each advancement seemed to herald a new era, with models like OpenAI's SORA, Meta's MovieGen, and others reshaping expectations for what AI could achieve in rendering visuals from textual descriptions.

OpenAI's SORA had gained traction as a sophisticated text-to-video model, building on the success of OpenAI's well-regarded language and image generation systems. With SORA Turbo, the latest iteration, OpenAI aimed to deliver not just higher efficiency but also greater fidelity in video rendering. However, despite these efforts, the model struggled to meet the increasingly high expectations of users who demanded lifelike outputs with precise adherence to prompts.

Meta, known for its focus on social connectivity, entered the fray with MovieGen, a tool designed to generate 1080p videos that appealed to creative professionals and enthusiasts alike. While its results were impressive in many scenarios, MovieGen fell short in areas such as user preference, fluid dynamics, and complex physics, leaving room for improvement. Other notable competitors, like Minx and Cing 1.5, found niches within specific creative circles but failed to dominate the broader market due to limitations in realism and coherence.

During this time, Google seemed to be trailing behind. Once the pioneer of AI breakthroughs, the company had been overshadowed by rivals who were faster to market with high-performing generative models. Critics speculated that Google's slower release cadence reflected a loss of its innovative edge, and the absence of a public launch for the original VEO model reinforced the perception that Google was falling behind. Yet, as history often proves, such assumptions can be premature.

VEO-2 marked a dramatic turning point. In a single release, Google managed to eclipse its competitors, delivering a model that set new benchmarks in generative AI. Unlike previous models, VEO-2 excelled in areas where others faltered—physics simulation, fluid dynamics, and long-form video consistency. The ability to generate videos that captured intricate details, such as the subtle jiggle of water in a glass or the exact speed of syrup flowing, showcased a level of sophistication that competitors had yet to achieve.

This moment signaled not just a comeback but a reinvention for Google. VEO-2 reestablished the company as a leader in AI innovation, proving that their deliberate approach to development had paid off. By outperforming industry giants and achieving results that many thought were years away, Google demonstrated that it was not only back in the race but potentially leading it once again.

Chapter 2: Unveiling VEO-2

VEO-2 represents a monumental leap in the evolution of text-to-video generation technology. Developed by Google as the second iteration of its video generation model, VEO-2 was engineered to overcome the limitations of existing models and redefine the standards of AI creativity. The technology behind VEO-2 combines advanced generative algorithms with a deeper understanding of physical interactions, allowing it to produce videos that are not only visually stunning but also remarkably coherent and realistic. Its purpose is clear: to set a new benchmark for how text can be transformed into high-quality, dynamic video outputs that mimic real-world physics and motion.

One of the most compelling aspects of VEO-2 lies in its performance on benchmark tests against other leading models in the field. OpenAI's SORA, a highly anticipated follow-up to its earlier generative systems, struggled to meet the expectations of a growing audience demanding precise adherence to prompts and flawless visual consistency. Meta's MovieGen, another key competitor, achieved respectable results in generating 1080p video but faltered when tasked with complex scenes or intricate details. Similarly, models like Minx and Cing 1.5, while popular in niche creative circles, could not match the overall capability of VEO-2 when it came to delivering fluid simulations and maintaining structural integrity in long-form outputs.

The numbers speak for themselves. Across various benchmarks, VEO-2 outperformed its competitors by a significant margin. In user preference tests, where participants evaluated videos generated by different models, VEO-2 was chosen nearly 60% of the time—an astonishing achievement in a field where subjective preferences often vary widely. By contrast, models like MovieGen and Minx managed only 30% preference rates, while SORA Turbo, despite its reputation, fell to the bottom of the rankings. This dominance extended to prompt adherence, with VEO-2 consistently generating outputs that closely aligned with user-provided descriptions, a critical factor in creative projects.

Breaking records wasn't limited to user preferences. VEO-2 also excelled in simulating complex physical phenomena, an area where other models often fell short. For instance, its ability to render liquids with near-perfect accuracy—capturing the subtle nuances of coffee being poured or syrup flowing in layers—demonstrated an unparalleled mastery of fluid dynamics. In other tests, it convincingly depicted intricate object interactions, such as a knife slicing through a tomato with visible vibrations and plump textures, further solidifying its place as a leader in generative video technology.

VEO-2's success is not just a testament to its technological sophistication but also a reflection of Google's commitment to pushing the boundaries of what AI can achieve. It redefines what users can expect from text-to-video generation, setting a new standard that competitors will find difficult to surpass. With VEO-2, Google has not only raised the bar but also solidified its position as the vanguard of the next wave of AI innovation.

Chapter 3: Physics Simulation Excellence

Creating realism in AI-generated video has long been one of the most formidable challenges for developers of generative models. Previous systems, despite their remarkable advances, often stumbled when tasked with replicating the intricate physics of the real world. The very nature of generative AI systems meant they frequently relied on approximations rather than true physical understanding, leading to outputs that, while visually impressive, lacked the subtle authenticity that makes motion and interaction appear natural. For example, liquid simulations were notoriously difficult, with earlier models struggling to render the chaotic yet structured flow of fluids. Similarly, object dynamics often fell short, with models producing results that felt static or, worse, unintentionally distorted.

VEO-2 represents a breakthrough in overcoming these limitations. One of its standout capabilities lies in its ability to simulate fluid motion with near-perfect accuracy. Consider the example of coffee being poured: VEO-2 captures every subtle nuance, from the splash and flow as the liquid meets the surface to the intricate ripples and reflections within the cup. Similarly, when rendering syrup, it accurately portrays the way it drips and layers, with the right speed and consistency, creating a strikingly realistic output. These achievements showcase VEO-2's deepened understanding of fluid dynamics, an area where many previous models faltered.

Object dynamics are another area where VEO-2 shines. When tasked with generating a scene of a knife slicing through a tomato, the model goes beyond merely depicting the action—it captures the essence of physical interaction. The knife's subtle vibrations as it cuts, the way the tomato's skin resists before yielding, and the plump texture of the sliced fruit all combine to create an output that feels lifelike. In another demonstration, VEO-2 showed its prowess in rendering water dynamics by simulating the slight jiggle of liquid in a glass as it was set down, adding yet another layer of realism that enhances the viewer's experience.

These advancements have profound implications for artists, creators, and industries reliant on AI-generated content. For filmmakers and animators, VEO-2 offers a tool that can bring their visions to life with unprecedented precision, reducing the need for costly and time-intensive physical simulations. Graphic designers and content creators can now explore ideas that were previously impractical due to the limitations of earlier models. Even educators and researchers stand to benefit, as the model's ability to replicate realistic scenarios opens doors for more engaging visualizations in fields like science and engineering.

VEO-2 is not just an improvement; it is a paradigm shift. By addressing the fundamental challenges of realism and delivering solutions that meet and exceed expectations, it empowers creators to explore their imaginations without being constrained by technical shortcomings. This leap forward signals a new era where AI tools are not merely functional but transformative, enabling a level of creative freedom that was once thought to be the exclusive domain of human ingenuity.

Chapter 4: Creative Capabilities of VEO-2

VEO-2's ability to generate creative outputs showcases not only its technological prowess but also its potential to revolutionize how we conceptualize and create visual content. The model's capability to handle wildly imaginative prompts with precision and coherence has opened doors for storytellers, animators, and artists to push the boundaries of their craft. Among the most intriguing examples are its outputs that blend humor, action, and cinematic brilliance, revealing just how versatile and capable this technology truly is.

In one memorable demonstration, VEO-2 generated a sitcom featuring potato protagonists. This whimsical example highlighted its ability to imbue inanimate objects with character and consistency, a feat that earlier models struggled to achieve. From scene to scene, the potato characters retained their distinctive features, even as they interacted, moved, and expressed emotions in a way that felt natural within the comedic framework. The result was a surprisingly engaging and coherent storyline that captured the charm and absurdity of the concept.

The model's versatility was further demonstrated through a high-speed car chase that culminated in the vehicle plunging through a waterfall. This sequence, rich in complexity, tested VEO-2's ability to maintain structural integrity and realism across a long-form action sequence. Unlike many earlier models that struggled with maintaining object consistency—often causing cars to morph into unrecognizable shapes—VEO-2 preserved the car's design and features throughout the chase. The physics of the scene, from the car's suspension over rugged terrain to the splash as it burst through the waterfall, were rendered with impeccable accuracy, adding a cinematic touch that felt both thrilling and authentic.

Another standout example was a kung fu scene featuring a potato hero clad in a flowing black leather coat, battling adversaries in the midst of heavy rain. The choreography of the fight, the dynamics of the rain-soaked environment, and the potato protagonist's consistent movements all demonstrated VEO-2's exceptional grasp of long-form coherence. Rain droplets fell in a way that interacted seamlessly with the character's movements, while the coat swayed realistically during every kick and punch. Such outputs would have been nearly impossible for earlier models to achieve without visible inconsistencies or glitches.

What makes these examples even more remarkable is VEO-2's ability to adhere to prompts with astonishing precision. Whether it's the texture of a rain-soaked leather coat, the exact timing of a waterfall splash, or the comedic expressions of potato characters, the model ensures that the core elements of the prompt remain intact throughout the video. This level of adherence is a critical advancement, as it allows creators to trust the model to deliver outputs that align closely with their visions, even in the most complex scenarios.

The significance of this prompt adherence extends beyond entertainment. For creators in fields as diverse as education, advertising, and research, VEO-2 provides a tool that can bring abstract ideas to life with unparalleled fidelity. It ensures that no matter how detailed or intricate the vision, the output remains coherent, consistent, and engaging, making it a truly revolutionary force in the world of generative AI.

Chapter 5: Understanding the Underlying Technology

Although Google has kept many specifics of VEO-2's architecture under wraps, the glimpses provided offer enough to reveal a model built on sophisticated engineering and innovation. At its core, VEO-2 operates on an advanced generative AI framework designed to interpret and translate textual prompts into high-quality, realistic video outputs. While previous models relied heavily on surface-level approximations, VEO-2 demonstrates a deeper "understanding" of the physical world, an achievement that sets it apart from its predecessors and competitors.

One of the defining aspects of VEO-2's framework is its ability to simulate complex interactions between objects and environments.

This stems from innovations in integrating physics-informed generative modeling, enabling the system to render actions and movements that follow the rules of reality. For instance, when depicting the cutting of a tomato, VEO-2 does more than recreate a visual representation—it captures the precise tension of the knife against the tomato's surface, the subtle vibrations of the blade, and the dynamic interplay as the slices fall onto each other. Such details suggest that the model's architecture incorporates nuanced layers of training data focused on real-world physics and object behavior, allowing it to produce results that feel authentic.

The model's capacity to handle complex simulations, such as fluid dynamics, further underscores its superiority. Unlike earlier systems that struggled to maintain coherence in liquid motion—often resulting in static or overly generalized representations—VEO-2 manages to deliver nuanced outputs. Whether it's the turbulent flow of coffee being poured or the layering of syrup on a surface, every motion is calculated and rendered with impeccable detail. This ability hints at an architecture designed to process and compute intricate variables in real time, ensuring that even the most unpredictable elements of physical behavior are accurately reflected in its videos.

What sets VEO-2 apart is not just the outputs it creates but the underlying innovation that powers its performance. Google has built on its foundational AI technologies, leveraging years of research and expertise in neural networks, natural language processing, and generative adversarial networks. This is evident in the model's ability to adhere to prompts with near-perfect precision. Unlike models that merely match keywords to visual templates, VEO-2 seems to "comprehend" the intent behind a prompt, enabling it to construct videos that align closely with user expectations. This depth of understanding is a hallmark of Google's approach to AI—prioritizing adaptability and coherence over mere aesthetic appeal.

Google's achievements with VEO-2 reflect a strategic commitment to innovation. Drawing from its foundational contributions to AI, such as the development of Transformer models, Google has created a system that not only surpasses existing benchmarks but also introduces new possibilities for generative applications. VEO-2's architecture integrates lessons learned from earlier experiments while incorporating breakthroughs that push the boundaries of what AI can achieve. By doing so, Google has reasserted its position as a leader in the field, proving that the deliberate pace of their research yields results that are both groundbreaking and transformative.

Ultimately, VEO-2 represents the culmination of years of progress in understanding how AI can bridge the gap between digital and physical realities. Its architecture, though not fully disclosed, is clearly built on a foundation of innovation and expertise, allowing it to achieve results that redefine what is possible in generative AI. This synthesis of technical brilliance and creative potential is what makes VEO-2 a true game changer in the world of artificial intelligence.

Chapter 6: The Industry Impact of VEO-2

VEO-2 marks Google's resounding return to the forefront of artificial intelligence, reclaiming a leadership position many believed it had relinquished. For years, Google's reputation as a pioneer in AI innovation seemed to waver as competitors like OpenAI and Meta made headlines with breakthrough models that captured the public's imagination. Yet, with the release of VEO-2, Google not only silenced its critics but also reestablished itself as a dominant force in the industry. By surpassing benchmarks and delivering capabilities that its rivals had yet to achieve, VEO-2 demonstrated that Google's methodical approach to research and development can yield results that redefine the field.

The success of VEO-2 has undoubtedly sent ripples through the industry, prompting a reevaluation among its competitors. OpenAI, known for its rapid advancements in natural language and generative AI, may need to rethink its strategy to stay competitive. SORA, OpenAI's text-to-video generation model, which had been hailed as a significant step forward, now pales in comparison to the versatility and realism offered by VEO-2. Similarly, Meta's MovieGen, despite its achievements in producing high-resolution video, struggles to match the physics simulations and prompt adherence that make VEO-2 stand out. These developments are likely to push competitors into a race to innovate, with new investments and collaborations aimed at closing the gap. The release of VEO-2 has not just raised the bar—it has shifted the industry's expectations of what is possible.

Beyond its immediate impact on the AI landscape, VEO-2 carries significant implications for various industries, particularly those rooted in creativity and media. For media production, VEO-2 offers a tool that can streamline workflows and reduce costs, enabling filmmakers and animators to generate complex scenes without the need for expensive physical simulations or extensive post-production work. Advertising, too, stands to be revolutionized as brands harness VEO-2's ability to create engaging, visually stunning content tailored to specific audiences with unparalleled speed and efficiency. Campaigns that once required weeks of planning and execution can now be conceptualized and rendered within hours, offering marketers a level of agility that was previously unimaginable.

The potential applications of VEO-2 extend even further, touching industries as diverse as gaming, education, and virtual reality. Game developers can use the model to generate immersive environments and lifelike interactions, while educators can create dynamic visual aids that make complex concepts more accessible. In virtual reality and augmented reality, VEO-2 could serve as the foundation for creating realistic, interactive experiences that blur the line between the digital and physical worlds. The implications of this technology are vast, promising to transform not just how content is created but also how it is consumed.

In reclaiming its leadership in AI, Google has sent a clear message: innovation is not a race to the fastest release but a commitment to achieving breakthroughs that resonate across industries. VEO-2 is more than just a technological marvel;

it is a testament to what is possible when ambition meets precision. By redefining the standards for generative AI, Google has not only solidified its dominance but also paved the way for a new era of creativity and innovation.

Chapter 7: Challenges and Limitations

While VEO-2 represents a groundbreaking advancement in text-to-video generation, it is not without its limitations. Despite its remarkable ability to simulate complex physics, maintain long-form coherence, and adhere to prompts with precision, certain areas still require refinement. One notable challenge lies in generating highly intricate textures or environments involving extreme levels of detail. For example, while VEO-2 excels in rendering realistic liquids and object interactions, it occasionally struggles with highly nuanced visuals, such as microscopic details or hyper-dynamic scenes involving dozens of interacting elements. These gaps, though minor in the grand scope of its capabilities, highlight the need for continued improvement to meet the most demanding creative requirements.

Another area that warrants attention is its handling of extremely abstract or metaphorical prompts. While VEO-2 demonstrates an impressive ability to interpret complex instructions, certain scenarios—particularly those requiring abstract representations or artistic reinterpretations—can lead to outputs that feel less cohesive or imaginative than intended. This suggests a potential limitation in its ability to "think outside the box," a quality that human creativity often brings to the table.

The question of public access to VEO-2 also raises significant concerns. As with any advanced AI tool, democratizing access presents both opportunities and risks. On one hand, making such technology widely available could empower creators from diverse backgrounds to experiment with and benefit from cutting-edge tools, leading to a surge in innovation across industries. On the other hand, the potential for misuse cannot be overlooked. The ability to generate realistic videos at scale could be exploited for disinformation campaigns, deepfakes, or other malicious purposes, posing challenges for ethical oversight and societal trust.

Google faces a delicate balance in determining how to distribute VEO-2. Offering controlled access through partnerships with established creators and organizations could mitigate misuse while ensuring that the tool is used responsibly. However, such restrictions might limit its transformative potential, leaving smaller creators or those in less resource-rich regions unable to access its benefits. Striking the right balance between accessibility and security will be critical as the technology matures.

Ethical considerations further underscore the need for vigilance in deploying VEO-2. As with any AI capable of generating realistic content, the question of responsibility looms large. Who is accountable if an output is used to mislead or harm? How should AI-generated content be labeled to ensure transparency? These are questions that must be addressed as such tools become more integrated into society. Moreover, the potential for these technologies to displace human creators in certain roles raises important discussions about the future of creative industries, the preservation of artistry, and the equitable distribution of economic benefits.

VEO-2 is undoubtedly a milestone in the evolution of AI, but it also serves as a reminder that with great technological power comes equally great responsibility. Addressing its remaining gaps, managing public access wisely, and navigating the ethical complexities of AI-generated content will determine not only the model's legacy but also the broader implications of this era of innovation. Google's commitment to advancing these discussions will play a crucial role in shaping the future of AI in a way that maximizes its potential while minimizing its risks.

Chapter 8: The Future of Text-to-Video Generation

Google's success with VEO-2 raises inevitable questions about what lies ahead. The model's achievements, while groundbreaking, are likely only the beginning of a broader strategy to push the boundaries of generative AI even further. Speculation about future iterations suggests that Google may aim to enhance VEO-2's capabilities in areas such as ultra-high-definition video generation, real-time rendering for interactive applications, and more nuanced handling of abstract prompts. Additionally, advancements in fine-tuning customization could empower users to align the model's outputs more closely with their unique creative visions, offering an unprecedented level of control over AI-generated content.

Another plausible direction for Google is the integration of VEO-2 with other AI systems. By combining the strengths of its image generation model, IM3, and the textual prowess of its natural language tools, Google could create an ecosystem where users seamlessly generate content across multiple media formats. This convergence of technologies would not only streamline workflows but also expand the creative possibilities for storytellers, educators, and innovators.

The implications of VEO-2 extend beyond its immediate capabilities, setting the stage for broader advancements in artificial intelligence. By demonstrating how a model can simulate realistic physics and maintain long-form coherence, Google has established a new standard for AI systems aspiring to emulate the complexities of the real world. This progress lays the groundwork for future breakthroughs in areas like augmented and virtual reality, where immersive environments demand an exceptional level of detail and interaction. Moreover, VEO-2's success could inspire further research into AI's potential to understand and replicate human-like perception, paving the way for applications that bridge the gap between human creativity and machine intelligence.

For creators, the opportunities presented by tools like VEO-2 are boundless. In storytelling, these advancements open doors to narratives that were once too costly or logistically complex to produce. Independent filmmakers and animators can now realize ambitious visions with the support of AI tools that handle intricate details and enhance production quality. In education, VEO-2 offers a new medium for engaging students, transforming dry textbook material into vivid, dynamic visualizations that bring concepts to life. Imagine science lessons featuring accurate simulations of natural phenomena or history classes that reconstruct pivotal moments in immersive video form.

Entertainment, too, is poised for a transformation. Game developers can use AI-generated videos to create rich, adaptive storylines, while musicians and artists can experiment with AI to produce innovative multimedia experiences. Even advertising and marketing professionals stand to benefit, leveraging VEO-2's capabilities to craft campaigns that resonate with audiences through highly personalized, visually captivating content.

The future of generative AI is one of collaboration between human ingenuity and machine precision. With VEO-2, Google has not only delivered a tool of remarkable power but also provided a glimpse into the possibilities that await when technology becomes a partner in creativity. As we look ahead, the potential for these tools to redefine entire industries feels boundless, heralding a new era where ideas that once seemed impossible are now just a prompt away.

Conclusion

VEO-2 stands as a shining example of how far artificial intelligence has come and the transformative potential it holds for the future. By surpassing its predecessors and competitors in realism, prompt adherence, and the seamless integration of physics and creativity, Google's latest text-to-video model has redefined the possibilities of generative AI. Whether it's rendering lifelike fluid dynamics, maintaining coherence across complex scenarios, or enabling entirely new forms of artistic expression, VEO-2 has set a new standard that others in the industry will strive to match.

The question of whether VEO-2 is a game changer or merely another step in the evolution of AI remains one for readers to consider. Its impact is undeniably significant, introducing tools and capabilities that can reshape industries ranging from filmmaking to education. Yet, like any groundbreaking technology, it comes with challenges, from ethical concerns about misuse to questions about accessibility and equity. These factors will ultimately determine whether VEO-2's legacy is one of transformation or a missed opportunity. However, based on its achievements so far, it's hard to view this model as anything less than a pivotal moment in the trajectory of AI.

As we close this exploration of VEO-2, one thing is clear: the future of AI is a story that is still being written. For those captivated by the intersection of technology and creativity, now is the time to stay engaged, informed, and curious. The implications of tools like VEO-2 extend beyond the technical into the realms of art, education, and society itself. Whether you're an aspiring creator, a seasoned professional, or simply an enthusiast, the advances in AI invite us all to reimagine what's possible.

So, as we move forward into this new era of AI-driven innovation, the question isn't just whether VEO-2 is a game changer—it's how we, as a global community, will use these tools to shape the world we want to create. Let curiosity guide you as we collectively navigate the exciting and uncertain future of artificial intelligence.

www.ingramcontent.com/pod-product-compliance
Lightning Source LLC
Chambersburg PA
CBHW070949220526
45471CB00007B/2960